EXOGRAVITATION:

Ongoing Revolution in Cosmology

Mystery Solved:
Dark Energy Gone, Dark Matter Redefined
Standard Model Fulfilled—Infinite Larger Structures

by

John T. Cullen

"I have not come to destroy the Standard Model, but to fulfill it."
(—JC)

Clocktower Books, San Diego

Exogravitation: Ongoing Revolution in Cosmology—Mystery Solved: Dark Matter & Dark Energy Gone, Standard Model Fulfilled, Ultimate Larger Structures Copyright © 2008-2021 by Jean-Thomas Cullen (a.k.a. John T. Cullen, John Argo). All Rights Reserved.

Visit Exogravitation (or 'Exograv') dot com.

The copyright certificate of registration in 2008, and deposit, were titled "Infinitesimal Godot Particles, Dark Matter, and Dark Energy: A Cosmological 'Conjecture of Everything'" by Jean-Thomas Cullen.

A first published edition appeared online in 2008 at Fictionwise.com as "Exogravitation, or Dark Energy is Dead—Crisis Among The Stars: Looming Storm in Cosmology" by John T. Cullen. A Second Edition of the same title appeared online through Smashwords in 2015. A Third Edition appeared via Amazon KDP in April 2020. This edition is a slightly revised version of the April 2020 edition. August 2020 Edition. Slight revisions vs. 2008, 2015, Apr 2020.

Print and E-book Editions at Amazon online. You can also order the latest print edition from your local bookstore.

You may not reprint or publish any portion of this book without written permission from the author and publisher. You may quote ("fair use") a sentence or two as needed when publishing your own comments, reviews, or other works, provided you quote exactly as published in this text, and attribute each use to author John T. Cullen citing the 2020 source (this work).

Clocktower Books, San Diego indie small press owned by John T. Cullen (sole proprietor), has been publishing online since 1996. No submissions at this time please (thank you).

Contact editorial@clocktowerbooks.com
Clocktower Books
P.O. Box 600-973 Grantville Station
San Diego, California 92160-0973.

Exhibit A: Massive Gravitational Object in Space (Galaxy)

Quick Lookup: https://en.wikipedia.org/wiki/Galaxy

"A galaxy is a gravitationally bound system of stars, stellar remnants, interstellar gas, dust, and dark matter.

"In 1995, the majestic spiral galaxy NGC 4414 was imaged by the Hubble Space Telescope as part of the HST Key Project on the Extragalactic Distance Scale."

CONTENTS

PRÉCIS 2020 7
NOMENCLATURE 9
1. GRAVITY CAUSES THE ACCELERATION 12
2. LARGER STRUCTURE(S) BEYOND OUR COSMOS 14
3. THERE IS STUFF IN THE MOTHERVERSE 15
4. RULES AND ASSUMPTIONS 17
5. GODOTS: NEITHER ENERGY NOR MATTER BUT A 'THIRD THING' 19
6. RANDOM BROWNIAN MOTION 22
7. ACCRETION SPHERE 23
8. CRITICAL EVENT C-1 26
9. CRITICAL EVENT C-2 27
10. CRITICAL EVENT C-3 (BIG BANG) 28
11. C-3 DARK 29
12. C-3 HEAVY 31
13. C-3 BRIGHT 34
14. COSMOPAUSE/ATTENUATION LIMIT 36
15. ETERNAL CYCLE 38
ABOUT JOHN T. CULLEN 40

Précis 2020

The germ of the Exogravitation idea struck me in the early 2000s while I was reading several books on a display in a San Diego bookstore. The display, as best I can recall today, nearly twenty years later, dealt with issues in current cosmology, and featured works by many of the leading cosmologists.

I had a moment of insight while reading about the enigma of our expanding, accelerating universe. An unknown force referred to by a cryptic title ('Dark Energy') causes the universe to not only continue expanding in all directions, at an ever-expanding rate.

This came as a surprise to astronomers and cosmology experts in the late Twentieth Century, who had expected their research to lead to one exclusive outcome from among three possible: (1) still expanding but ever more slowly due to a central or aggregate gravitational force; or (2) past the ultimate expansion, no longer even slowly expanding, but actually contracting toward (probably) an ultimate collapse maybe back to some infinitesimal dead point; or, (3) the Fred Hoyle solution from the 1930s, a universe in stasis (steady state; neither expanding nor collapsing). The outcome was a totally unexpected, inexplicable discovery: (4) our cosmos is not only expanding, but at an ever increasing rate. How can that be?

Dark Energy seems to defy the laws of Conservation of Matter and Energy. We are confronted with a mystery. It's established that there is this force, or dark energy, so where has it been hiding? Matter and energy cannot be created *ex nihilo*, from nothing. The term Dark Energy is what engineers call a 'black box,' meaning an unknown or an anticipated solution when the inputs and outputs of a process are known or defined.

At that moment in the bookstore, I was struck by the germ of an idea. The notion of an ever-accelerating force baffled me, as it does everyone else, but there is one common vector in the known universe that will consistently cause such a result: gravitation.

In simplest terms, the metaphoric apple is hanging on a tree (potential energy). A gust of wind causes the stem to snap, leaving the apple in momentary free-fall. The centripetal attraction of earth's mass (gravitation) pulls the apple (a tiny gravitational mass in itself) toward the earth's surface. The apple begins to move, faster and faster, until the surface stops its motion.

That sort of vector motion seemed to me to very closely mirror the accelerating expansion of the cosmos.

Then it struck me: what if the universe is not being pushed to expand ever faster by some hidden internal force that defies the laws of conservation? What if the universe is instead being pulled apart (falling in all directions) by a vast exterior gravitational attractor? That became my thought experiment (*Gedanken-experiment*) that I call exogravitation.

Call it a theory, call it a hypothesis; I am content to call it either or those, or simply a thought experiment.

The exogravitation idea implies many things, among them the existence of a larger universe, most likely a Cantorian infinity of spaces and eternities of time, in which our cosmos is simply one grain of sand on an endless shoal of space and time.

In the 1920s era, Edwin Hubble and other researchers demonstrated a number of things: that the heliocentric theory of Copernicus (and ancient predecessors) is forever gone, because not only is the sun not the center of all things, but our solar system is not at the center of the universe, and our galaxy (on whose remote rim our solar system is an obscure neighborhood) is not the entire universe but itself a mere grain of sand on a vast beach of galaxies. Today we face yet another paradigm shift, to borrow a term from Thomas S. Kuhn (1962 book The Structure of Scientific Revolutions). The 1920s work of Edwin Hubble, compounded with that of Albert Einstein, Georges Le Maître, caused a paradigm shift beyond the Copernican view, and just as importantly, made us realize that our galaxy is just one of many. Just as there are structures in the universe larger than our galaxy, so now we can seriously entertain the notion that there are infinitely many universes like our own.

Exogravitation poses yet another hyperstructure: a super-cosmos of infinitely many universes, or eternities of time and infinities of space, as many scientists have long suspected.

The Hubble group demonstrated that the Milky Way is not the entire universe, but one of myriad galaxies. In fact there are yet larger structures (e.g. walls of galaxies). With the Exograv theory, we can presently only suggest one larger structure; but who is to say there may not be 'walls of universes?'

Nomenclature

We'll use some ad hoc terminology for now.

Exogravitation: the simple, universal laws of gravitation applies to larger structures; we may also use the term 'exograv' for short;

Motherverse: for lack of an existing term, that would be the ultimate structure containing infinitely many universes, and possibly larger structures yet to be determined.

Note: I use the terms universe and cosmos interchangeably.

Gravitation Elements: the dark matter or sub-sub-particles that are common throughout the motherverse.

At times I may refer to these gravitational elements as go-dots or godots (not from Samuel Beckett's 1948/9 play Waiting For Godot, but from my 2003 SF novels Mars the Divine and Orwell in Orbit 2084, in my Empire of Time SF series, where I first started processing cosmological ideas; the term godot is my own little nickname for the round game-pieces in the Japanese game of Go).

These godots do not possess attributes of 'matter' or 'energy.' Godots are infinitesimal, even in comparison with sub-atomic particles. Godots cannot really be called particles, nor energy; they only possess one characteristic, which is of a minute gravitational charge or value. This attraction-force causes them, in random Brownian Motion (analogous to molecules and small debris in a pond), to clump together in larger and larger units. In the following thesis, I call these accretion spheres (successively C-1, C-2, and C-3).

C-2 and C-3 embody the so-called Big Bang of the Standard Model. The letter C stands for 'Critical.' There are three major critical stages in this expanded, more complete and understandable Standard Model. The so-called Big Bang no longer comes uniquely out of nowhere, but has a standard before, during, and after phase as I will explain. The Big Bang (still a good term) is now part of a standard, explainable mechanism that happens infinitely many times across eternities of space in the motherverse.

Accretion Spheres. As already mentioned, the vast spherical shapes (universes, in the Motherverse) form out of dark matter (gravitation elements, or godots). These godots accrete by random Brownian Motion until sufficiently dense and heavy. At the C-1 critical phase, a sufficiently massive accretion sphere collapses along lines of the Standard Model to explode in a Big Bang that initiates a universe. These spheres go through three moments of change that I will describe (see the text) as C-1, C-2, and C-3 transitions, where C-

3 is the Big Bang moment.

C-1 is the initial accretion sphere when enough gravitational elements have clumped together.

When the overall gravitational force of C-1 is robust enough to cause an inward collapse, the sphere shrinks suddenly and *implosively* to a tiny, compressed sphere. This is C-2, a moment that cannot last because the forces of implosion vectoring from all directions toward a single focal point are unsustainable or irresistible. C-2 explodes outward in all directions from the combined energy of the C-1 collapse and the C-2 compression, causing Big Bang C-3.

C-3 is the final creative moment when the forces of implosion reverse in a massive *explosion* of unimaginable power, spewing visible and invisible debris in all directions (a Big Bang).

Output will include massive spheres of clumped godot material (black holes) pulling visible matter and energy along to spin and form standard galaxies, as well as stray bits of intermediate material, and thirdly, loose gravitational element debris we have found in the form of 'dark matter.'

Cosmopause: This is the out limit and the end of a universe's life, the limit of expansion and attenuation. By analogy think of the solar-pause, that theoretical sphere at which the aggregate gravitational force of the solar system loses its ability to act upon an escaping object.

Summing up: a typical cosmos, like our own, goes through a life cycle that I describe, from a Critical (1-2-3) beginning (Big Bang is part of it) toward our currently observed stage of accelerating expansion. The universe is constantly attenuating as it expands, until it reaches a point I call the Cosmopause. At the Cosmopause, the universe disintegrates (better: vanishes) like a bubble, meaning that its foundational dark matter is all that is left, but not as an integral gravitational entity.

The residual gravitational matter (godot raw material) then drifts randomly in the larger motherverse, amid similar debris of infinitely many other universes. Some of these eventually clump to form other universes and thus endlessly reiterating the life cycle.

* * * *

So again, as observed in my bookstore moment, each universe is pulled apart overall evenly by the ambient gravitation of the overall Motherverse. A universe 'falls in all directions.'

As a general tendency, because of the enormous gathering

power of each accretion sphere (which eventually collapses and initiates a Big Bang starting a new universe), and because of the explosive power of each Big Bang, it is unlikely that individual universes can form adjacent to each other. Right there, we have an instance of how it is possible to begin forming some suppositions about this nearly unknowable motherverse of which we are a part.

I suspect that the ambient gravitation, leading to accelerated expansion and pulling our cosmos apart, should tend to be fairly uniform overall.

Only recently, 2020, I have become aware of the work of Thomas Kuhn (1922-1996) who, in his 1962 philosophical book The Structure of Scientific Revolutions, independently echoes some of my own observations about how scientific ideas evolve over time and through structures of authority, consensus (including peer review), and acceptance (or not). My observations in the historical area are not meant to call attention to themselves per se. I am primarily interested in communicating my idea which I call Exogravitation; but I could not help in my investigations to learn some broad outlines of discovery that beg for notice.

I am eliminating a lengthy history of cosmology in this article (to appear at www.exograv.com) so we can proceed to the core idea (Exograv or Exogravitation) but context is important, and therefore worth a few quick notes. And, I might add, my article over a decade ago, leading to the chapter on an exograv revolution in cosmological ideas, was part of my own learning process.

I close this chapter with the observation that:

"I come, not to destroy the Standard Model, but to fulfill it."

1. Gravity Causes The Acceleration

I suggest that the acceleration of the expansion of our universe is not caused by a mysterious dark (energy) push-force at all. Instead, it is caused by the pull-force gravitation (a commonly understood and accepted mechanism rather than some mysterious black box concept).

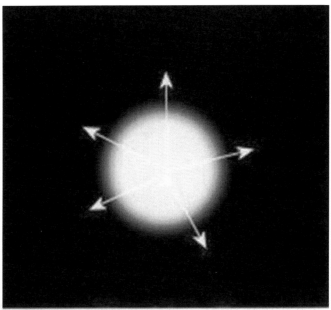

Fig. 1: Expanding Universe

We have known for a long time that gravitation is a fundamental force in the universe. Recently, experimental evidence has shown that the universe is not steady state (never changing size), nor is its acceleration slowing in such a manner that its aggregate mass will eventually cause it to collapse. That left one other possibility, as scientists saw it—namely that the universe is accelerating fast enough to overcome its aggregate gravitation, and will simply continue expanding forever.

It has turned out that none of these scenarios is true, though the last one comes close. It turns out that not only is the universe flying apart, fast enough to overcome its gravitational mass, but in fact it is accelerating due to some mysterious force. Working within the Standard Model for a single, unitary universe, scientists have suggested a 'dark energy' that is pushing the universe to expand faster and faster.

This 'dark energy' is largely undefined, but seems to imply some really remarkable departures from standard science. If this overwhelmingly powerful energy is real, then where has it been hidden? What is it? Nobody knows, because it is unknowable, and it is that because it does not exist.

Instead, I suggest, the universe is being pulled apart in all directions. Just as Newton's legendary apple fell faster and faster, the longer it was in free-fall before hitting the ground, so the universe is flying apart faster and faster—in all directions.

2. Larger Structure(s) Beyond Our Cosmos

If our universe is being pulled apart in all directions, that implies a larger mass all around the universe, and inherently some sort of structure in the sense that humans think of solar systems and galaxies as 'structures.'

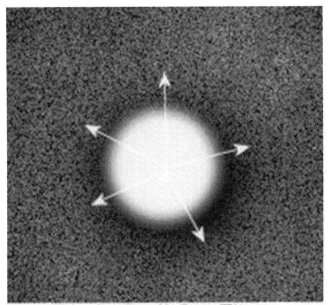

Fig. 2: Stuff Out There

This is not, in itself, as novel an idea as one may initially suppose. It does not contradict existing science—which, in fact, supports it, based on a major precedent.

This hypothesis suggests a further iteration of the Hubble paradigm shift, which confirmed that our galaxy is not the whole universe. Just as there are a vast number of galaxies outside our own, comprising a newly defined and much more vast universe—so I am suggesting that there is a much larger structure beyond our universe, in which our universe is just one of many, perhaps infinitely many. I am not the first to suggest it, though I seem to be the first to suggest that the accelerating expansion is due to gravity, which irrevocably requires larger structures beyond what we call the universe.

3. There Is Stuff In The Motherverse

If the meta-gravitational hypothesis is assumed to be true, it follows without recourse that there is a larger structure—which may have its own substructures and superstructures with which we need not concern ourselves now.

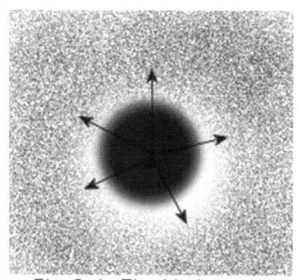

Fig. 3: In The Motherverse
= random Brownian motion - godot gravitational platelets =

What can we know about this conjectural motherverse? Surprisingly, we can construct a viable overall model as I am doing in this article. As more researchers look at it, and test concepts experimentally, there will no doubt be modifications as is natural. I would not presume to claim to have the final structural model.

For one thing, it has to be bigger than any daughter universe, including our own. The aggregate gravitational mass of the motherverse is far greater than the mass of our universe, by a significant factor, since it is capable of pulling any universe apart as it is doing to ours right now.

The motherverse, greatest force in nature, has the aggregate gravitational force of all collective universes, plus all the free-floating godots that have yet to make it into an accretion sphere.

The aggregate mass is evenly distributed, causing our universe to expand uniformly in all directions.

A special case to the contrary would be that there is some large substructure (a galaxy of universes, by a poor analogy for reasons to be understood shortly) nearby that would pull our universe more in one direction than any other.

In the simplest scenario, assume an evenly distributed, granular motherverse that pulls equally and gradually in all directions.

4. Rules And Assumptions

That brings up three points of procedure. (1) As much as possible, we restrain our sequiturs to the simplest case possible. We will conjecture only the least complicated hypothetical larger structure, called the motherverse. (2) We assume the laws of science do not vary under any circumstances; and any emergent circumstances would only imply there are laws we do not yet understand. (3) for simplicity, we assume that space is the same inside and outside our universe, and that all universes are simply local effects. Each is of limited duration. Each is bounded locally, temporarily, and dynamically by its own cosmopause, which expands with ever greater velocity, and represents the outer boundary of the coherent mass/energy of the whole.

As mentioned elsewhere, it is not entirely clear yet if we should even think of these other structures as separate universes, or simply manifestations within an infinitely large universe that has undergone the type of paradigm shift like that introduced with the much expanded Hubble universe in the early 20th Century.

Fig. 4: A Go-Dot (pure grav)

In simplest terms, our galaxy is still part of the known cosmos. The parts are integral to the whole. Most cogent is this question: is all the godot material (dark matter) in our universe the result of C-3, or can there be stray amounts of random godot 'stuff' drifting around that is part of the motherverse, but didn't undergo our unique C-3?

Does material from other universes or structures blend with ours, especially the majority godot platelet material that is, by analogy, the common 'hydrogen' of the unseeable motherverse. Another way to state this generally is to ask if there is an underlying foundational universe composed of godots, in which C-3 universes briefly flare in and out of existence (over eons and vast spaces by our subjective measures); or if each universe is an entirely discrete volume that is hermetically sealed and self-contained. I have no answers for these questions, except to say that we must cautiously and systematically pry apart our assumptions to make way for possibly truer insights.

Simplicity dominates in what is essentially an Einsteinian or Gaussian thought experiment. Per Occam's Razor, we want to stick with the simplest answer or construct at every turn. Given any choice between a simpler alternative and a more complex one, we will want to pick the simpler. The less stuff we 'make up,' the better, since this is a series of conjectures based on the first conjecture, that the expansion of the universe is caused by gravity. At the same time, we cannot slam the door on unexpected twists or complexities. This research will likely take generations to achieve further significant insights.

We are at the beginning of a clean, fresh paradigm. It does not discard the old (e.g., Standard Model), any more than Hubble's universe of trillions of galaxies did away with the laws of a single galaxy mistakenly seen as the entire cosmos.

5. Godots:
Neither Energy nor Matter but a 'Third Thing'

If gravity is causing the accelerating expansion, then gravity is our dark energy—no longer dark, no longer mysterious.

Fig. 5: Infinitely Many Go-Dots

As mentioned in previous pages, there is stuff out here. But what sort of stuff? In sticking to the rule of simplicity, I'm going to just name one kind of stuff, and say it is neither matter nor energy, but the most fundamental thing of all things.

As discussed at length in my side notes (previous page), the go-dot is neither a wave or a particle, neither energy nor matter, but a third thing. Since the only property we know for sure it has is gravity, I'll stick with the bare bones of necessity and call it a platelet of pure gravity. It is the fundamental and self-defined unit of pure gravitation.

* * * *

NOTE: Here is a slight digression, inherited from earlier editions. In the current edition, I skip over the history (and my own learning process at the time) and go directly to the exogravitation theory that is, after all, the critical core of this endeavor. Worth noting: in my 2008 edition, I elaborate on how existing scientific models keep being patched as new, contradictory data arrive...until a point of no return arrives, where we have to ditch the old model and move to something radically new. More info at www.exogravitation.com (in work for general readers) and www.exograv.com (online for invited reviewers).

* * * *

Earlier Editions. Of the examples I cite in the 2008 edition, the most jarring is that of Ptolemy's 2nd Century concept that the universe consists of *planetae* (wandering objects) that move along circular paths in the heavens (in a geocentric universe). When in early modern times it was discovered that Mars often moves in apparent retrograde fashion as observed relative to Jupiter, Ptolemaic theory was patched up as 'cycles and epicycles.' This added smaller, retrograde cycles to account for the apparent motion of planets, until the entire Ptolemaic myth had to be discarded in the age of Copernicus, Kepler, Newton, Tycho Brahe, and other observers on the way toward a more modern theory that included the development of mathematical calculus describing elliptical orbits in a heliocentric rather than geocentric universe.

Sistine Chapel. Other examples of paradigm shift cited in my earlier (2008, 2015) editions included the Catholic Church's response to the influx of Hellenic, Turkish, and Arabic scholarship versus the predominance of European Medieval Latin and Scholasticism. This is captured poignantly, evocatively, and perhaps desperately in the purpose and layout of the Sistine Chapel, named for Pope Sixtus IV (late 1400s, right at the cusp of the Early Modern age). The Sistine Chapel, whose complex parts travel from Creation at the entrance across all of time to the Last Judgment above the main altar, tries to reconcile new understandings of how the Christian message interlocks with its spiritual sources in Hebrew Scripture and polytheistic ('pagan') cosmology, the latter as personified in the ancient cult of Sibyls.

Einstein/Le Maître/Hubble/et al. While the earliest paradigm shift(s) mentioned date to Classical Athens (where geocentric and heliocentric notions already sparred amid accusations of blasphemy against polytheistic deities like Zeus (Theos), and I discuss the Sistine Chapel in its very transitional historic age at the culmination of the Renaissance and beginning of Early Modern times, and I already discussed the collapse of Ptolemaic 'cycles and epicycle' theory, likewise the 1920s became a locus of jarring change. The last vestiges of Copernican heliocentric theory fell by the wayside, even as our concept of the universe broadened (vastly) so that the Milky Way galaxy was no longer seen as the entire universe. In fact, even in our galaxy, our solar system is not even close to the galactic center, but in fact resides in an obscure trailer park on the remote fringes of one galactic arm filled with stars (and, we now know) full of solar systems.

Another key take-away is that we have to be prepared to find ever larger structures, as exemplified by the discovery of 'walls of galaxies' and the like. At Palomar Observatory in San Diego County, California, as one enters the main building, one sees inscribed over the doorway a number with at least twenty zeroes after it, that describes however vaguely the unimaginably vast number of stars in the universe.

Exogravitation adds yet another, larger, unimaginably vast structural layer (which I call the Motherverse, but some will call it the Metaverse or the like). Our cosmos or universe (I tend to use the two terms interchangeably) is just one of infinitely many universes much like our own whose multi-billion year life span from birth (Big Bang, Standard Model) to death (attenuation beyond its internal gravitational or centripctal reach or Cosmopause) is just a blink of an eye in cosmic time.

6. Random Brownian Motion

For the Motherverse not to collapse on itself, the infinitely many go-dots remain apart or discrete. Each is so tiny that we have no way of measuring it. In effect, the go-dot is the irreducible unit of dark matter, whose only measurable effect is gravitation in large clumps.

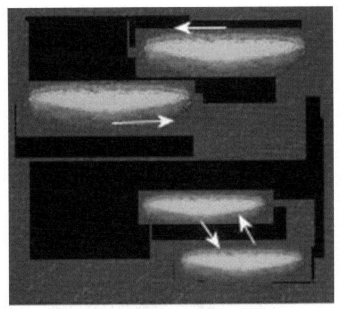

Fig. 6: Drifting, Clumping
= random Brownian motion =

Go-dots have a tiny unitary gravitational strength (not wave, not matter). By moving about (Random Brownian Motion), they come into contact with each other for what necessarily follows.

All those endless tiny godots aren't standing still. Like tiny specks or molecules in a pond, they are in constant motion very similar to Random Brownian Motion in fluidics theory. The godot is the fundamental 'thing,' platelet, of existence. We'll now learn how the go-dot become basic building blocks of any universe. Since space is to be considered the same inside and outside the event of a typical universe, go-dots exist both inside and outside universes. As we'll see, their behavior is differentiated within the critical stages of making and unmaking a typical universe.

7. Accretion Sphere

Let us look at the underlying process of star formation, since we posit that the laws of science are the same everywhere. Stars form when random matter drifts close and mutually attracts. Nature works in brute force, minimal-effort ways (like water seeking its proverbial lowest common level). When sufficient matter clumps together (80 or more Jupiters) we get a full-blown star, whose mass causes the reduction of hydrogen into a hot, formless plasma that gives off light and heat, as helium forms in the compression. As it happens, something like 80% of elemental matter in our universe is hydrogen. This fits neatly and predictably, since hydrogen is the simplest of the bright or visible elements. Next, much of the rest helium, with all else nearly an afterthought (we'll discuss this later in our go-dot process). Even the simplest element, and its irreducible unit or quark, is gigantic compared to the go-dots (dark matter, pure gravity) floating freely about. The gravitational sink created by a large entity like a star causes a distortion in the distribution of dark matter. Apparently (another subject for experiment or observation) the centrifugal pressure of light and heat emitted by the star more than offsets the centripetal clumping tendency of floating go-dots.

In the larger motherverse, the same type of process applies. When go-dots start clumping, they form an accretion sphere, whose shape is naturally dictated by the democracy of attraction. All units strive with equal success toward the core of the gravitational sink created by the nascent universe, which is as yet simply an accretion sphere of go-dots.

As more and more go-dots clump together, their combined and mutual attraction starts creating a gravity sink that 'vacuums' more and more go-dots on an ever-faster and more massive scale.

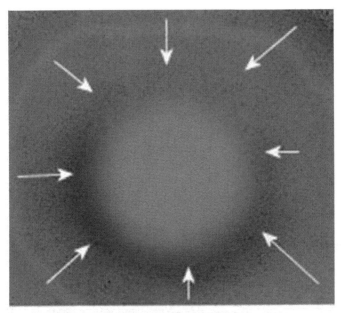

Fig. 7: Accretion Sphere

This is analogous to star formation, but on a much larger scale with much smaller discrete units. The accretion sphere is not yet a universe, but a sort of reverse-universe, like a sock that has to be turned inside out before one can wear it. The turning-inside out of the sock, by analogy, is the Big Bang moment in the Standard Model.

Infinitely many godots drift around the motherverse in random Brownian motion. Each godot is so small, and its gravitation identity so tiny, that any two godots have to be really close to one another to mutually stick. But stick they shall, under the right circumstances. In the aggregate, either discretely, in clumps, in accretion spheres, and within finished universes, godots are the fundamental element of the overall gravitation of the motherverse, which is the most powerful force in nature.

There is, logically, utter inevitability about this process. As more and more godots randomly cling together (just as the far more massive hydrogen molecules find each other inside the universe in empty space), accretion disks start forming. Once an accretion disk starts forming amid the essentially random and uniform distribution of go-dots in the motherverse, the process accelerates incrementally. The process takes on a life and inevitability of its own, and rushes faster and faster to its predictable conclusion as it 'vacuums' every available go-dot unto itself from an increasingly large surrounding volume of the motherverse.

Unimaginable but finite numbers of these very tiny go-dots, like the finest dust beyond imagination, end up building an entire universe—ours, and infinitely many, in an eternal cycle of accretion, explosion, and destructive attenuation.

The accretion sphere itself drifts around in the motherverse, slowly propelled by the aggregation of random kinetic energies of its component until at some point maybe it loses its random Brownian motion. It becomes a fixed point in the aggregate gravitation of the motherverse. However, more and more godots come drifting by.

Could we see this happening? For a time I thought not, but now I see no reason for there not to be at least the distant glow of functioning universes. So maybe there is a dim *Gegenschein* of some sort. We won't see the godots, but it's conceivable (a minor point here) that the accretion sphere may begin to emit some sort of energy, perhaps from all the accumulated kinetic energy of the random Brownian motion that now has nowhere to go. This should be detectable in the residual turbulence of the early universe, and herein may lie some opportunities for experimentation and observation that my expert requested.

As the accretion sphere continues growing, it becomes a space-time gravity sink in its own right, and starts vacuuming surrounding space clean of godots.

In fact, vacuum is a casual word used in the previous sentence, but it raises an interesting point—is the empty vacuum surrounding the accretion disk somehow translated into the universe that's about to form? I have no idea.

The accretion sphere grows and grows, until by some as yet unknown limits, it reaches the first of three critical events.

8. Critical Event C-1

The accretion sphere builds with increasing speed, converting the inrushing kinetic energy of go-dots into potential energy at the core. This core energy builds until some (unknown) critical point, at which the forces within the core can no longer hold their equilibrium, and something must give.

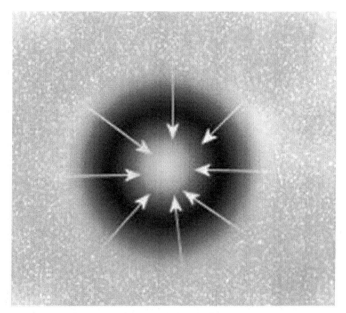

Fig. 8: C-1 Collapse

At this point, the accretion sphere collapses on itself, under the gravitational hyper-pressure of its myriad go-dots and their combined potential energy, which continues to build at an enormous velocity and force. The nature of the material of the sphere changes from unattached godots to a kind of primordial energy composed of pure gravity and potential energy, like a hyper-plasma. That is critical event C-1.

9. Critical Event C-2

Instantly, the collapsed C-1 contracts to an inflection point that could conceivably be the same as Lemaître's primordial atom.

This is *the moment before* the Big Bang.

The material changes to a form of ultra-compressed plasma that we can only metaphorically describe as a sort of godot concrete or platelet titanium. A total energy transfer happens instantaneously, from totally potential to totally kinetic (the natural state of the go-dots before their accretion). This single moment is Critical Event C-2.

Fig. 9: C-2 Compression

In C-2, the accretion sphere here is compressed to a dot. Time and space (to become our specific universe, not all times and spaces) are utterly compressed like the most massive black hole imaginable. This is an unsustainable condition that changes just as instantaneously as it formed.

10. Critical Event C-3 (Big Bang)

Instantly, the C-2 material now shatters. We could say it explodes, but as it turns out that is too smooth and simple to describe the complex series of events that now follow. This is the instant when our metaphorical sock turns inside out. All that has been stored up in the compressed C-2 flies apart with C-3, creating our universe as we know it. Local space-time starts expanding explosively, and with it come all the circumstances engendered in the Standard Model. This is a high-energy period of great violence and perturbance, which eventually emerges as the universe we feel that we know.

Fig. 10: C-3 Big Bang

Three types of material fly forth, as the core material shatters. It does not uniformly explode, as the observational and measurable results of the known universe show. We'll call this material C-3, and classify its component types as (1) Dark, (2) Heavy, or (3) Bright. '

11. C-3 Dark

Most of the C-3 mass comes apart and flies out as godots in the metaphor of an expanding raisin cake. It's the dark stuff we detect only indirectly by its gravitational effects. This is our so-called dark matter. This is a fine dust made of godots, pure gravity platelets. It is the same stuff that went into the accretion sphere, having lost its gravitational, attractive stickiness. It is the same within the cosmopause, and beyond the cosmopause, of any universe. It interacts with the other material thus: (11 C-3 Dark, 12 C-3 Heavy, and 13 C-3 Visible or Bright):

After the initial boiling phase, the universe begins to form into Dark, Heavy, and Bright matter. In Figure 11, we illustrate the tremendous forces by which the standard galaxy unit forms with a so-called black hole (heavy fragment of accretion disk still intact) at the center, and swirls of invisible go-dot platelets (dark matter) swirling around it as well as the intermediate fine C-3 material that starts 'unpacking' as visible light and measurable mass manifestations. This accounts for the fact that vast amounts of invisible (Dark and Heavy) material constitute most of a galaxy's mass, whereas the Bright or Visible material is a small fraction (est. 5%).

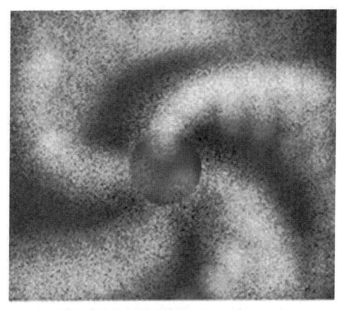

Fig. 11: Visible Universe

12. C-3 Heavy

I make an important distinction between two forms of Dark Matter resulting from Critical Event Three (C-3). One is C-3 Dark, consisting of random go-dot or gravitation platelets that spread with relative uniformity across empty space within the cosmos (inside the limit of the cosmopause that can be mathematically projected as a theoretical sphere, analogous to the heliopause in our solar system). The other is C-3 Heavy.

C-3 Heavy is a subset of the more general C-3 Dark. The difference is that the specific subset C-3 Heavy will be found as black hole material (which is C-3 Dark stuff resulting from the C-3 critical event or 'big bang' moment.

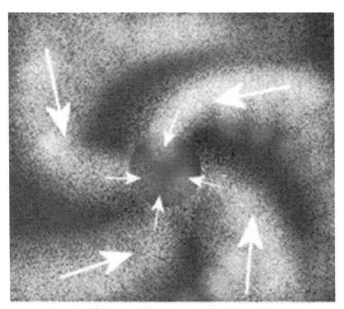

Fig. 12: Galaxy Formation

Of the Dark Matter ejected from the C-3 point (Big Bang), significant amounts will remain stuck together by their natural platelet gravitational attraction. Remember, these are neither particles nor energy waves, but pure primordial 'Third Thing' somethings I have dubbed go-dots. Because they are neither matter nor energy in commonly measured forms (but 'Third Thing' go-dot gravitation platelets), I see the following divisions of 'stuff' in our universe through a somewhat different lens. This does not destroy the Standard Model but (I think) sharpens it. Fundamental thing to realize: our terminology is driven by our observational history, like the old Ptolemaic cycles and epicycles. Inherent in all or most existing models is the simple distinction between 'visible' and 'not visible.' Basically, 5% of the stuff in our universe is 'visible' (detectable by eye or telescope or the like) while 95% is 'invisible,' as noticed initially in the orbital dynamics of observed galaxies: namely, that the amount visible as starlight and light waves could not contain sufficient gravitational mass to hold the galaxy together at the velocity at which it is rotating around its central core (which we now know typically is a super-gravitational locus or 'black hole').

The Standard Model (quick look-up at Wikipedia) breaks down the three outcomes very similarly: "Dark matter constitutes about 26.5% of the mass-energy density of the universe. The remaining 4.9% comprises all ordinary matter observed as atoms, chemical elements, gas and plasma, the stuff of which visible planets, stars and galaxies are made. The great majority of ordinary matter in the universe is unseen, since visible stars and gas inside galaxies and clusters account for less than 10% of the ordinary matter contribution to the mass-energy density of the universe." (sources at Wikipedia article). I don't know to what extent my C-3 Dark and C-3 Heavy (both 'dark matter') might be analogous (or not) with such existing theories as 'cold dark matter' and 'hot dark matter.'

We can agree on the (rounding) 5% visible matter (C-3 Bright). A divergence comes in that there is no so-called Dark Energy (black box terminology) in the Exogravitation Model. Rather, the inclusive super-set Dark 95% consist of the core C-3 Dark scattered through our local cosmos *including* the subset C-3 Dark (Heavy) that concentrates into black holes. By the dynamics of the early universe immediately after C-3 (Big Bang event or moment) and the resulting outward rush of material in a turbulent chaos resolved into logically gravitation-induced gyres, the result is galaxy formation. While there are various structures in the cosmos, there appears to be a commonality of rotating disk-shaped galaxies, often with so-called arms or spokes wheeling in balance around that central black hole core.

These vast blobs of go-dots remain stuck together and go spinning away from the C-3 (Big Bang) point in all directions. As they fly out (after the initial 300MY condensate phase identified by scientists for the early universe) their enormous dark mass gathers C-3 Bright material to form what we call galaxies ('milks'... talk about nomenclature!).

13. C-3 Bright

A small amount of the C-3 material, maybe 4%, moves away in all directions from the primordial point that was the original C-2 universe. That is the pinpoint predicated in the Big Bang of the Standard Model. This stuff is an intermediate result that is not lightly bound enough to loss all clumping and resolve into individual go-dot or dark matter. At the same time, it is not dense enough to become the Heavy C-3 that forms galactic nuclei with their black holes and quasars.

Fig. 13. Above photo (credit NASA Hubble Deep Field team 1996) shows the visible universe (C-3 Bright) matter and energy.

In a common large galaxy, the dark hole (heavy C-3 that remains tightly clumped from the C-2 implosion) forms the center of a nascent galaxy. It creates the galaxy by drawing dynamic, curving streamers of Bright C-3 (the elements from Hydrogen on up; visible matter and energy). This bright C-3 represents an intermediate stage of C-2/3 'unpacking' to ultimately resolve back into C-3 Dark matter (primordial go-dots).

In effect, this Bright C-3 is unpacking itself on the way to returning to the primordial state of go-dots. expands into the universe that is now unpacking itself. This visible material has a lot of energy, and wraps itself around the black holes that attract large swaths of it.

This suggests the mechanism that describes precisely what is observed in galaxies that appear to have far more invisible material than visible. The three types of C-3 material form the black hole (Heavy), the swirl of visible stars, planets, dust, and other elemental material (Bright), and the invisible mass of platelets (Dark).

The baryonic matter and other forms of particles, along with the 92 elements and all, at their most fundamental stage consist of godot platelets stuck together as part of a lingering C-3 effect.

NOTE: I am often compelled to think of the overall mechanics of the motherverse as a kind of endless 'breathing' in and out, creating universes and absorbing (destroying) them. We should be able statistically and mathematically to estimate 'breathing' periods although there is not one lung or pair of lungs at work (all metaphor) but infinitely many such pulsating processes. They will tend to occur far apart, because the accretion process will tend to vacuum up stray godot material into mutually exclusive gravitational zones or pauses.

14. Cosmopause/Attenuation Limit

The motherverse is immediately at work, with its vast gravitation, pulling the new universe apart. This process will take a finite time, measurable in billions of years. It should be possible to calibrate all events in these eternal cycles into predictable quanta, part of time scales mentioned in the note at end of the previous section.

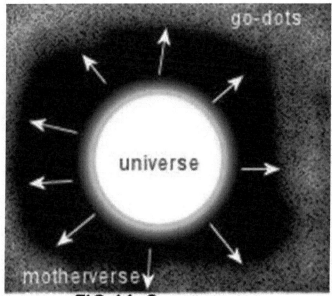

FIG.14: Cosmopause
(black is inside universe, gray out in Motherverse)

At the outer edge all around an expanding universe is what we may call the cosmopause (edge of white sphere, fine aqua line, false color). The cosmopause is a limit inside the motherverse, beyond which our universe has not yet grown. As our universe expands (constantly accelerating), it follows the expected sequence of events just outlined. Gradually, it attenuates to a point where the empty space of the motherverse, with its drifting godots, cannot be meaningfully distinguished from what is left of our universe.

One of the many black box (unsolved) questions in all of this is whether a given universe can potentially push aside material from other universes, or if they can overlap as the primordial Big Bang energy (C-3) of a universe gradually diminishes to the point that, as its attenuation limit (cosmopause) spreads beyond centripetal limits, material from adjoining universes with higher centrifugal forces (to use simple physics terminology) begins to mingle with the material of the dying universe in question.

15. Eternal Cycle

At this point, we're back to godots, a sort of invisible, immeasurably fine 'dust' of pure gravitation platelets, and the cycle is complete. It is may be possible that an observer standing outside in the motherverse could see many universes. They might look like a field of stars or smudgy galaxies. Time and space are relative within this framework, and everything works as predicted by existing science, to the extent that existing science has wrapped itself around this larger structure in the new paradigm.

FIG.15: Cosmic Cycle Repeats

= infinite range eternal cycle beginningless endless loop =

All C-3 ejecta (Dark, Heavy, Bright) in a universe eventually degenerate to godot material in the motherverse, after the end of a cosmopause. This forms seed material for new accretion spheres in an unending cycle. We are, as the song says, not only 'star dust, we are golden,' but we are the material from which infinitely many new universes form in an eternal motherverse of no beginning or end.

So there we are, having portrayed the cycle of evolutionary, natural, non-designed creation through all of its phases.

We have taken the path of simplicity at every turn. More

complicated scenarios might involve what happens when multiple accretion spheres form near each other (they most likely jam together, and form a single new universe).

It seems unlikely that universes drift in the motherverse, because their mass would be a large attractor versus the greater aggregate mass of the motherverse, and would tend to stay fixed if the motherverse is pulling equally from all sides. That idea, however, does raise a tantalizing possibility.

The question arises whether there are larger structures (of universes) in the motherverse, as there are galaxies, great walls of galaxies, and other vast structures in our universe. I can only suggest that it is possible, given what we know at present.

Thank you for your time and interest.

JTC
San Diego, California

About John T. Cullen

For furhter information related to this theory, please visit https://www.exogravitation.com

Visit the author's personal website (and multi-site weplex) at www.johntcullen.com or his indie small press publishing house Clocktower Books in San Diego www.clocktowerbooks.com, the world's first publisher of true e-books in 1996. That means (please note) the first proprietary rather than public domain, full-length e-book novels published directly online for reading in HTML or download in TXT formats rather than on portable media like CD-ROM). Innovations at the time included publishing novels online (HTML) in sequential weekly chapters, typically on Sunday afternoons (San Diego or PST) to be read by avid fans around the world. Readers at the time tended to be tech-savvy, often highly educated, and working in a world of revolutionary new ideas. There were still relatively few home computers and no hand-held processors like today, so most readers came to these John T. Cullen articles and novels holding their morning tea or coffee, in offices around the world including (I still have many of their rave e-mails) USA, Canada, Australia, New Zealand, U.K., Germany, France, South Africa, Taiwan, and other points around the globe.

* * * *

John T. Cullen has been a professional writer, researcher, journalist, editor, and publisher for over half a century. As a journalist, he is also a science and history writer. More information about the author at his website (https://www.johntcullen.com) and his webplex of thematically linked websites including Clocktower Books (https://www.clocktowerbooks.com) and the publisher's museum site reflecting a quarter century online (https://www.museum.fyi).

The author holds three university degrees, including a B.A. in Liberal Arts (University of Connecticut); a B.B.A. in Computer Information Systems (National University); and an M.S. in Business Administration (Boston University).

Nonfiction includes:
Dead Move: Kate Morgan and the Haunting Mystery of Coronado (1892 true crime and later famous ghost legend near San Diego).
Sator Enigma: Ancient Roman Mystery Solved
A Walk in Ancient Rome (first authorized edition due 2022)

Fiction includes:
CON2: The Generals of October (political thriller)
Siberian Girl (historical suspense)
Lethal Journey (fictional 1892 noir dramatization based on *Dead Move* above)
The Christmas Clock (Dark Seasonal Fantasy praised in a personal Jan 2008 rave fan mail letter from Ray Bradbury)…

More info at the webplex.

Made in the USA
Middletown, DE
08 April 2022